AF510269

The AI Empire: Building a Smarter World Through Systems and Vision

The AI Empire: Building a Smarter World Through Systems and Vision

Michael Harden

Michael Harden

The AI Empire: Building a Smarter World Through Systems and Vision

All rights reserved
Copyright © 2026 by Michael Harden

No part of this publication may be reproduced, distributed, or transmitted in any form or by any means, including photocopying, recording, or other electronic or mechanical methods, without the prior written permission of the publisher, except in the case of brief quotations embodied in critical reviews and certain other noncommercial uses permitted by copyright law.

Published by Spines
ISBN: 979-8-90222-859-2

Contents

Introduction: The Builders of the New Empire 7

Preface: From Vision to System 11

1. The Age of Intelligent Systems 15
2. The Empire Mindset 19
3. Foundations of AI Civilization 25
4. The Systems Architect 31
5. Building DGIX — The Case for Vision-Driven Technology 39
6. Law, Ethics, and the AI Governance Frontier 47
7. The Human Element — Emotional Intelligence in a Machine World 55
8. From Data to Destiny — The Economics of AI Empires 63
9. The Legacy Equation — Vision × Systems = Immortality 71
10. The Future We Are Building 77

Epilogue 85

Appendix A: Influences and Recommended Reading 87

Introduction: The Builders of the New Empire

Empires were once built from stone, steel, and sweat. Today, they are built from code, data, and vision.

The world is entering an age unlike any before it — an age where intelligence itself has become infrastructure. Artificial intelligence is no longer a tool or a trend; it is the foundation of a new civilization. What steam was to the 19th century, and electricity was to the 20th, AI will be to the 21st: the invisible force powering everything we touch, create, and become.

But technology alone does not build empires. **Systems and vision do.**

Without systems, innovation collapses under its own weight. Without vision, systems become machinery without meaning. The great challenge of our era — and the great opportunity — is to unify both.

When I founded **Disney Group International Exchange (DGIX)**, I didn't set out to create another technology company. I set out to build a system — a framework for how humans and machines could work together intelligently, ethically, and exponentially. DGIX was never just about software; it was about sovereignty — the ability to architect your own destiny in a digital world dominated by algorithms.

The AI Empire is not a story about domination or control. It is about **design** — the deliberate construction of systems that serve humanity, elevate efficiency, and extend the boundaries of what we believe possible. This empire has no borders, no armies, and no flags. Its territory is **data**, its economy is **innovation**, and its citizens are the creators who dare to think differently.

We are witnessing the rise of digital nations — enterprises and ecosystems that move faster, think deeper, and operate with the precision of machine logic and the intuition of human leadership. These are the new empires of our time: Amazon's logistics intelligence, Oracle's cloud architecture, NVIDIA's AI factories, and the countless startups building the future from garage to global.

The question isn't whether artificial intelligence will reshape our world — it already has. The question is **who will shape the systems that shape us.**

This book is a roadmap for the builders, the dreamers, the architects of tomorrow. It's written for the visionary leaders who understand that the next generation of power won't come from owning assets, but from orchestrating **intelligence** — human and artificial — into harmonious systems. It's a blueprint for how to think like an empire builder in a digital age, guided not by conquest, but by creation.

We are all standing at the threshold of a vast frontier.
Those who master systems will master destiny.
Those who lead with vision will define civilization itself.

The future belongs to the architects of intelligence.
This is their manual.
This is The AI Empire.

Preface: From Vision to System

When I first started my journey in technology, I wasn't chasing artificial intelligence — I was chasing understanding. I wanted to understand how systems worked: financial systems, legal systems, corporate systems, and later, digital systems. What fascinated me most wasn't the machinery itself, but the invisible architecture behind it — the logic that made it all function.

Over time, I realized something simple yet profound:

Every great empire in history was, at its core, a **system of systems** — political, economic, military, and cultural. And in our time, **AI is the newest system** humanity must learn to govern wisely.

I wrote The AI Empire because I believe we are living through the most consequential transformation since the Industrial Revolution. The power once concentrated in monarchs and manufacturing now resides in algorithms and infrastructure. And just as Rome mastered roads, and

Britain mastered ships, the next great builders will master **data, code, and cloud intelligence**.

My background as both a lawyer and technologist has shown me two sides of progress — the side that builds and the side that regulates. As an attorney, I've seen how laws attempt to catch up with innovation. As a technologist, I've seen how innovation can transcend law. The balance between those two is where civilization either advances or collapses.

That's why I founded **Disney Group International Exchange (DGIX)** — not just as a company, but as a construct. A model for how to integrate governance, intelligence, and vision into one functioning ecosystem. We're building technology that thinks, systems that scale, and solutions that align profit with principle. In many ways, DGIX is both the laboratory and the living case study behind this book.

I am a student of philosophy as much as I am of code.
I often return to the teachings of **Marcus Aurelius**, who reminded us that "The universe is changing; our life is what our thoughts make it." AI is simply the next manifestation of that truth. It's not here to replace us — it's here to reveal us.
To reveal how we think, how we lead, how we decide what is right and wrong when machines are capable of thinking for themselves.

This book is not written for data scientists or lawyers alone. It's written for **leaders** — those who believe in the power of systems, and those who understand that real leadership is about designing order from chaos. It's written for entrepreneurs who want to scale intelligently, for policymakers who want to legislate wisely, and for thinkers who understand that the next empire won't be built on territory, but on **trust** and **technology**.

The AI Empire is my contribution to a growing conversation about how we build responsibly while dreaming audaciously.

It's a call to every visionary — from Silicon Valley to Easton Town Center — to design not just products, but **principles**; not just companies, but **civilizations.**

We are no longer living in the Information Age.

We are living in the **Age of Intelligence** — and the question is no longer whether we will build an empire, but what kind of empire it will be.

— **Michael Harden** Founder, CTO & General Counsel
Disney Group International Exchange (DGIX)
Columbus, Ohio

Chapter 1
The Age of Intelligent Systems

The first age of humanity was survival.
The second was industry.
The third — the one we are now entering — is intelligence.

For the first time in history, our inventions are beginning to think for themselves. Machines are not just repeating our commands; they are interpreting them, refining them, and sometimes improving upon them. We have moved from programming logic to programming **learning** — from automating work to automating wisdom.

This is the dawn of the **Age of Intelligent Systems** — an era where code becomes cognition, and infrastructure becomes alive.

The Shift: From Power to Perception

In the Industrial Age, power was measured in horsepower, oil barrels, and tons of steel.

In the Digital Age, power is measured in **intelligence per second.**

Every business, government, and institution is being redefined by one central question:

How intelligently can you operate your system?

The new economy isn't defined by access to capital — it's defined by access to computation. The companies that will dominate the next fifty years will not simply build products; they will build **learning systems** that evolve faster than human bureaucracy can adapt.

That's why Amazon doesn't just sell products — it predicts demand.
Why Oracle doesn't just store data — it structures intelligence.
Why NVIDIA doesn't just build chips — it builds thinking factories.

Each of these companies has understood something fundamental: in the modern world, architecture is destiny.

The System Mind

To build effectively in this new age, we must think in systems, not silos.

Systems thinking requires a shift in perspective — from the individual part to the interconnected whole. It's how nature operates, how empires endured, and how modern technology scales.

Every successful civilization in history — from the Roman Empire to the digital empires of today — thrived because it mastered **systems integration**. Roads, trade, and law were not separate elements; they were components of a cohesive framework that produced stability, efficiency, and growth.

In the same way, AI systems today must integrate across boundaries: data, ethics, governance, and humanity. That is what distinguishes innovation from civilization.

The DGIX Philosophy

When I founded **Disney Group International Exchange (DGIX)**, it was never about chasing the next technology trend — it was about creating a **systematic philosophy of intelligence.**

DGIX operates on a simple premise: systems create scale, and scale creates sustainability. Our work focuses on designing frameworks where AI enhances human potential, not replaces it; where automation liberates creativity instead of limiting it; and where intelligence becomes a shared resource, not a corporate monopoly.

At DGIX, every project — from Oracle ERP integration to AI-driven governance — begins with a single question:

"How can this system make people smarter, not just systems faster?"

That's what separates true vision from technical obsession.
Technology without philosophy is machinery.
Philosophy without technology is poetry.
But when you merge the two — you create an empire.

A New Kind of Empire

The empires of the past conquered geography.
The empires of the future will conquer complexity.

They will be built not by generals or monarchs, but by thinkers, engineers, and leaders who understand that the greatest territory to map is the human mind — and the algorithms that extend it.

In the coming chapters, we will explore how these systems are being built, how they're governed, and how visionary leaders can shape them responsibly.Because make no mistake — the future will belong to those who understand **systems**, and who have the **vision** to align them with purpose.

The Age of Intelligent Systems has begun.The question is no longer whether we will build an AI Empire — but **what kind of empire it will be.**

Chapter 2
The Empire Mindset

Every empire begins as an idea — a vision in the mind of a builder who refuses to accept the limits of the present.
It's not resources that build empires. It's resolve.
It's not technology that sustains them. It's thinking.

The true empire builder does not chase trends. He creates frameworks.

He sees patterns where others see chaos, systems where others see noise.

He is not content to lead projects; he builds platforms.

He is not chasing scale; he is designing sovereignty.

That is the Empire Mindset — the discipline of seeing the world not as it is, but as it could be, through the eyes of systems and vision.

1. The Philosophy of Builders

History remembers leaders like Augustus, Genghis Khan, and Alexander the Great for their conquests — but their true genius wasn't warfare; it was **organization.** They didn't just take land — they built systems of control, commerce, and culture that lasted centuries.

Rome's roads were not just infrastructure; they were information networks. Persia's couriers were not just messengers; they were data pipelines. Egypt's hieroglyphs were not art; they were data storage systems carved in stone.

Every enduring empire was built on structure.

And today, the empires being built in Silicon Valley, Shenzhen, and Seoul follow the same principle — but their bricks are data, their roads are fiber optics, and their armies are algorithms.

2. Vision as Infrastructure

Vision, in leadership, is not poetry — it's architecture.

It's the mental blueprint that gives shape to unseen potential.

The empire mindset requires seeing every decision as a structural investment — not a transaction. Whether you're designing an AI model, building a company, or shaping a national strategy, your vision must operate like infrastructure: durable, scalable, and adaptable.

That's why some companies last a decade, and others last a century.

They build for the unseen.

At **Disney Group International Exchange (DGIX)**, our approach to vision mirrors that philosophy. We don't ask, "What can AI do today?"
We ask, "What system must exist so AI can evolve tomorrow?"

That question transforms your organization from reactive to sovereign — from a startup into an empire of innovation.

3. The Discipline of Systems

To think like an emperor is to think in systems.

Empires thrive because they understand cause and effect at scale — logistics, culture, communication, governance. They are, by nature, multilayered organisms where everything is interconnected.

In technology, the same principle applies.
When you design systems — whether it's an AI supply chain, an ERP platform, or a data governance framework — your true mastery lies in **integration**.
The empire mindset doesn't obsess over parts; it orchestrates wholes.

That's why the next generation of leaders must learn to be **systems architects**, not just managers. To lead AI transformation, you must understand how technology, ethics, human behavior, and law converge. You must design for complexity, not fear it.

4. Stoicism and Empire

Empire-building is not for the impatient.
It requires Stoic discipline — the ability to act strategically while accepting what is beyond control.

Marcus Aurelius, the philosopher-emperor, ruled Rome not merely through power, but through principle. His strength was his restraint. His vision was sustained not by ambition alone, but by alignment with reason.

In the age of AI, Stoicism is more relevant than ever.
The leader who cannot master his emotions will never master his algorithms.
The empire mindset is calm under pressure, clear under chaos, and committed under uncertainty.

The Stoic ruler and the modern CTO share the same creed:

Control the controllable. Anticipate the variables.
Build systems that outlast emotion.

5. The DGIX Method: Building from Vision Down

At DGIX, we apply the empire mindset by building from vision down, not from task up.

Every system begins with a **philosophical anchor** — a principle that defines why it exists.
From there, we architect the **framework** — the processes, technologies, and rules that make it real.

Finally, we execute the **operations** — the daily activities that keep the system alive.

Vision → Framework → Execution.

That is the DGIX Hierarchy of Systems Thinking.

This method ensures that no product, platform, or initiative exists in isolation. Everything is part of the greater ecosystem — the empire of intelligence we are consciously designing.

6. The Psychology of Sovereignty

At its core, empire-building is about **sovereignty** — the ability to act without permission.

In a world governed by algorithms, true sovereignty comes from understanding the systems you depend on.

The empire mindset refuses dependency.
It seeks autonomy through architecture — whether that means owning your data, controlling your cloud infrastructure, or designing your own digital laws.

Sovereignty is not just a geopolitical concept anymore. It's digital self-determination.The leaders who understand this will not just build companies — they will shape civilizations.

7. Beyond Leadership — Toward Legacy

The empire mindset looks beyond profit to permanence.

It is not satisfied with quarterly gains or viral moments. It is obsessed with legacy systems — ideas and infrastructures that outlive their creators.

That is the purpose of The AI Empire: to remind us that technology must serve a moral architecture; that leadership must extend beyond tenure; and that every system we build today becomes the civilization our children inherit.

We are not just writing code.
We are writing history.

The empire mindset demands vision anchored in virtue, strategy rooted in systems, and leadership driven by legacy.

Because empires don't fall when their armies weaken — they fall when their vision fades.

The next chapter of humanity will be written by those who think not in terms of projects, but in **paradigms.**

Not in terms of goals, but in **governance.**

Not in terms of profit, but in **permanence.**

That is the Empire Mindset.
And it is the foundation upon which The AI Empire must rise.

Chapter 3
Foundations of AI Civilization

Every civilization stands on pillars.
Egypt had its pyramids and priesthood.
Rome had its roads and laws.
America had its Constitution and markets.
Our civilization — the one being born in the 21st century — rests upon three invisible but powerful foundations: **Data, Cloud, and Connectivity.**

These are the arteries of the new world.

They carry not water or blood, but information. And whoever understands how to organize, govern, and protect them will shape the destiny of humanity's next empire — the **AI Civilization.**

1. Data: The New Natural Resource

If oil fueled the 20th century, data fuels the 21st.
But unlike oil, data doesn't deplete — it multiplies. Every

click, transaction, image, and word becomes part of a global nervous system constantly sensing, learning, and predicting.

Yet data alone is not power — interpreted data is.

In the same way that raw ore is useless without metallurgy, raw data is meaningless without intelligence. The modern leader's task is not to collect information, but to transform it into **insightful action.**

At **DGIX**, we treat data as **sovereign infrastructure.** It's not just a business asset — it's a civic one. Data defines rights, governs systems, and underpins every decision from national security to personal finance.

In an AI-driven world, those who control their data control their destiny.

That is why data governance — privacy, ethics, and accountability — will be to the AI age what constitutions were to democracy: the **blueprint of power.**

2. Cloud: The New Infrastructure of Empire

Every empire has required infrastructure.
Rome had roads; Britain had ships; America had railroads and the internet.
The new empire runs on **cloud infrastructure** — the invisible scaffolding of modern civilization.

Cloud is not just a storage solution. It is the **foundation of global collaboration, computation, and creation.** It is where intelligence lives, scales, and evolves.

The cloud is the **city** of the AI civilization — always awake, borderless, and self-expanding.

The companies that dominate the cloud — Oracle, Microsoft, Amazon, Google — are the new architects of empire. Their datacenters are our cathedrals. Their APIs are our trade routes. Their uptime is our oxygen.

But with great infrastructure comes great imbalance.

Power concentrated in a few global entities can lead to **digital imperialism** — where nations depend on corporations for the very continuity of their economies.

The challenge for the next generation of leaders is to **decentralize resilience** — to design systems where intelligence is distributed, secure, and sovereign.
At DGIX, our vision aligns with this mission: to make intelligence accessible, equitable, and accountable.

3. Connectivity: The Lifeblood of Intelligence

If data is the resource and cloud is the infrastructure, **connectivity** is the bloodstream.
Connectivity is not just Wi-Fi and 5G; it's the continuous dialogue between every human, machine, and system on the planet.

The power of connection is exponential.
One node adds value, but a network creates civilization.
The internet began as a tool for communication — it has evolved into the nervous system of humanity. AI is what gives that system **consciousness.**

In this interconnected ecosystem, the boundaries between the physical and digital worlds are dissolving. Every home, car, city, and corporation is part of a living, breathing system that learns in real time.

And the question is no longer whether we are connected, but how intelligently that connection serves us.

4. Law and the Algorithmic Constitution

No civilization survives without law.
The Roman Empire had the Twelve Tables.
The United States had its Constitution.
The AI Civilization will require something equally fundamental — a **Digital Constitution.**

As algorithms become lawmakers in their own right — deciding what content we see, how credit is granted, or who receives medical attention — we must establish **ethical boundaries and governance systems** that align with human values.

As General Counsel of DGIX, I believe the next century will belong to leaders who can speak both legal language and machine language. Those who can encode ethics into algorithms will define the future of justice.

We are approaching a moment where law must become **machine-readable** — and morality must become **machine-teachable.** That is the frontier of digital governance.

5. The Economic Engine of Intelligence

Civilizations thrive when they harness energy efficiently.
The Industrial Revolution gave us mechanical energy.
The AI Revolution gives us cognitive energy.

AI's true power lies not in replacing labor, but in amplifying intelligence.
Economies built on human labor will always hit limits.
Economies built on machine learning can scale infinitely — if guided by ethics and architecture.

In the AI Civilization, **productivity becomes perception.** The faster your systems can learn and adapt, the more value you create.

Wealth will no longer come from what you own, but from what your systems know.

6. Civilization by Design

Civilizations do not emerge by accident — they are architected. And the builders of the AI Civilization are not emperors or engineers alone; they are philosophers, policymakers, and system thinkers.

The same way ancient builders designed aqueducts to deliver water, we must design data streams that deliver wisdom.
The same way ancient lawmakers crafted codes to protect citizens, we must craft digital laws to protect human dignity.
The same way nations once built walls, we must now build

ethical firewalls that defend against corruption of intelligence.

In every generation, the question of civilization is the same:

"How do we manage power?"

The only difference now is that our power thinks.

7. Toward a Smarter World

The foundations of the AI Civilization are not just technical — they are **moral, philosophical, and structural.** To build a smarter world, we must build smarter systems. Systems that are transparent in design, purposeful in intent, and scalable in service.

That is the mission of The AI Empire: to build systems that serve humanity, not enslave it; that enhance creativity, not extract it; that preserve wisdom, not just generate data.

We are not the first civilization to rise — but we may be the first to rise with our machines.

The empires of the past were measured by territory.
The empires of the future will be measured by **intelligence per person.**

Our responsibility — as builders, as leaders, as visionaries — is to ensure that the AI Civilization becomes not a monument to machinery, but a masterpiece of human design.

Chapter 4
The Systems Architect

Every great civilization has its builders.

But only a few understand that to build is not to assemble — it is to **architect.**

An architect is not a mere designer.
He is a strategist of structure, a curator of complexity, and a guardian of coherence.
He sees the unseen — the way systems breathe, evolve, and respond to pressure.
And in the age of artificial intelligence, **the Systems Architect** is the new philosopher-king.

1. The Modern Architect

The ancient world had its architects of stone.
The industrial world had its architects of steel.
The digital world has its architects of systems.

In this age, the Systems Architect stands at the intersection of **vision, engineering, law, and human behavior.** He must understand how to build infrastructures that not only scale, but also govern themselves.

He is part strategist, part scientist, part statesman.

And his greatest tools are not hammers or blueprints — but **frameworks, logic, and ethics.**

At **Disney Group International Exchange (DGIX)**, this role defines our leadership model.
We do not simply write code; we write constitutions for systems.
We design architectures that ensure every layer — human, algorithmic, legal, and operational — aligns toward one intelligent outcome: progress with purpose.

2. Systems Thinking as Leadership

To lead in the AI era, one must think in systems.
Traditional leadership manages people.
System leadership manages **patterns** — cause and effect, feedback loops, incentives, and behaviors.

A Systems Architect asks different questions:

- How does this process scale under pressure?
- How do small decisions ripple through large structures?
- What happens when the system meets entropy — failure, bias, or corruption?

These are not technical questions alone — they are governance questions. That is why the modern CTO must be part engineer and part ethicist.

Because when systems fail, they rarely fail mechanically — they fail morally.

3. Architecture as Ethics

Architecture is the hidden morality of the modern world.
Every algorithm, every workflow, every cloud environment encodes values — what to prioritize, what to ignore, who to include, and who to exclude.

If your architecture lacks integrity, your empire will collapse under its own complexity.

That's why ethical architecture is not optional — it's existential.

As both a technologist and an attorney, I've seen this truth firsthand:
Regulation tries to correct what architecture failed to anticipate.
So the Systems Architect's responsibility is to **design morality into infrastructure.**

When we embed transparency into data models, we embed trust into institutions.
When we design explainability into AI, we preserve accountability in governance.
When we align profit with principle, we sustain civilization.

The architect's first blueprint is not code — it is conscience.

4. The DGIX Framework: Vision → Structure → Scale

At DGIX, every solution begins with a framework.

We call it the **VSS Model — Vision, Structure, Scale.**

1. **Vision:** Define the purpose — the "why" behind the system.
    - Without purpose, efficiency becomes exploitation.
2. **Structure:** Design the architecture — the "how" that gives the vision form.
    - Structure turns purpose into process.
3. **Scale:** Build for sustainability — the "what next" that ensures the system grows intelligently.
    - Scale turns process into permanence.
4. This model allows DGIX to deliver complex integrations — from Oracle Cloud ERP to AI-driven governance tools — while maintaining clarity and cohesion.
    - It's how we transform technical projects into living systems.

5. The Law of Interconnectedness

Every decision inside a system affects the whole.
A single misaligned process can distort an entire network.
That's why the Systems Architect must see beyond silos — to the ecosystem.

The best leaders don't fix problems; they fix patterns.
They design systems so that integrity is the default, not the exception.
This is how you future-proof not just a business, but a civilization.

Rome endured for centuries because its systems — military, legal, and economic — were interconnected.
Modern enterprises endure for the same reason.
AI, cloud, and law are not separate disciplines; they are **layers of one structure.**

6. The Human Core of Systems

Amid all the technology, one truth remains: systems are built by, for, and around people.

AI can process; only humans can perceive. AI can optimize; only humans can interpret. AI can mimic wisdom; only humans can live it.

The Systems Architect never forgets this.
He designs with empathy — understanding that every interface, automation, and algorithm ultimately touches a life.
Human-centered architecture is not a luxury. It is the soul of sustainability.

As leaders, our mission is not to replace humanity, but to refine it through design.
To use technology as a mirror — one that reflects who we are and who we aspire to be.

7. Building Immortality Through Design

Systems outlive their creators.
That is the quiet reward — and burden — of the architect.
What you design today will govern tomorrow.

Every database schema, every governance policy, every line of code becomes part of the digital sediment that future generations will inherit.

You are not just building infrastructure — you are building memory.

The Roman aqueducts still stand because they were built with precision and principle.
So too must our AI architectures — designed not for quarterly performance, but for generational endurance.

That is the legacy of the Systems Architect:
He builds not for applause, but for eternity.

8. The Architect's Oath

If doctors have the Hippocratic Oath, engineers must now have the **Architect's Oath:**

> I will design with intention.
> I will align intelligence with integrity.
> I will build systems that serve humanity, not dominate it.
> I will measure progress not by profit alone, but by permanence and purpose.

> I will remember that every system I design becomes someone's world.

This is the ethos of The AI Empire. And this is the code by which every modern architect must live.

We are no longer designing software — we are designing society.

And the question that will define our era is not how intelligent our machines become, but how **responsibly we architect the intelligence that shapes them.**

The empire is rising — line by line, code by code, system by system.

And its architects are already among us.

Chapter 5
Building DGIX — The Case for Vision-Driven Technology

Every empire begins with a founding idea — a single sentence that captures the will to build something greater than oneself.

For **Disney Group International Exchange**, that idea was simple:
Technology must serve vision — not the other way around.

When I founded DGIX, I wasn't building just another systems-integration firm.

I was building a **philosophy in motion** — an organization that would merge the precision of engineering with the conscience of law, the scale of cloud computing with the soul of human creativity.

DGIX was born out of a conviction that systems, when designed with integrity, can become instruments of progress. That conviction has guided every product we've shipped, every partnership we've formed, and every decision we've made.

1. The Genesis of a Vision

The inspiration for DGIX came during a period of reflection on the intersection of law, technology, and legacy.
I saw a widening gap: technologists were building systems faster than policymakers could understand them.
Corporations were automating without accountability.
Governments were regulating without comprehension.

I realized that what the world needed wasn't another IT contractor — it needed a **systems interpreter.** A company that could speak three languages fluently: technology, governance, and vision.

That's how DGIX began — not as a startup, but as a **strategic response to disorder.** A structure designed to bring coherence to complexity.

2. The DGIX Framework: Vision → System → Scale

Every DGIX engagement begins with one fundamental question:

What vision are we building toward?

If the vision isn't clear, the technology will eventually fail.

So, we reverse the traditional approach: instead of leading with tools, we lead with **truth** — the "why."

Once the vision is defined, we translate it into **systems** — frameworks, architectures, and governance models that give it shape.

Only then do we pursue **scale**, ensuring that every process can grow intelligently without compromising ethics or purpose.

This model — Vision → System → Scale — has become the DGIX DNA.
It keeps our teams aligned, our clients accountable, and our mission intact.

3. From Projects to Ecosystems

In the early days, DGIX delivered Oracle Cloud ERP implementations, AI automation, and IT modernization for mid-market enterprises.

But as our clients evolved, so did we.

We began to see each engagement not as a project, but as part of a larger **ecosystem of intelligence.** When we connected an AI-driven HR system to an organization's compliance platform, it wasn't just a technical integration — it was **digital governance.** When we optimized supply-chain analytics, it wasn't just efficiency — it was **economic architecture.**

We realized DGIX was doing more than building solutions.

We were helping organizations **govern themselves through technology.**

4. Law as Architecture

My legal background has shaped DGIX as much as my technical one.
In both law and code, structure determines behavior.
A contract is a system of obligations.
An algorithm is a system of instructions.
Both require clarity, logic, and foresight.

That's why DGIX treats compliance and governance not as constraints, but as **design principles.** Our legal lens allows us to anticipate risks before they manifest — to encode accountability directly into architecture.

We don't just integrate systems; we **institutionalize integrity.**

5. The Culture of Builders

DGIX's culture is built on three principles: **discipline, design, and dignity.**

- **Discipline** keeps us strategic — every decision must serve the long game.
- **Design** keeps us intentional — every solution must align with purpose.
- **Dignity** keeps us human — every innovation must elevate people, not diminish them.

Our teams are small but sovereign — thinkers, coders, and counselors united by one mission: to build responsibly.

In an industry obsessed with speed, DGIX values **stability.** We don't just chase innovation; we **govern** it.

6. Partnerships and the New Consortium Model

Empires rise through alliances, not isolation.

DGIX's partnerships with major cloud providers, AI vendors, and legal innovators reflect that truth.

But unlike traditional vendor relationships, ours are **consortiums of vision.**

Each partner must align with our ethical framework: transparent AI, secure data handling, and human-centered design.
We choose alignment over convenience — because empire-level systems require empire-level principles.

7. Building for the Next Generation

DGIX is more than a company — it is a **platform for successors.** We train young professionals not just in technology, but in philosophy.
They study Marcus Aurelius alongside Python.
They learn that a balanced algorithm requires a balanced mind.
We are not simply producing consultants — we are cultivating architects of the AI civilization.

That is how you build a company that becomes a movement:
you align your systems with your story.

8. The Empire in Motion

Today, DGIX stands as a model of what I call **Vision-Driven Technology** — technology that exists to express a philosophy, not just fulfill a function.
We are integrating Oracle ERP with blockchain, deploying AI governance models, and helping organizations design digital constitutions that guide their own growth.

But our true product is not software.

It's sovereignty. It's the ability for enterprises — and eventually, nations — to govern their digital futures with intelligence, integrity, and intention.

DGIX is not an empire of servers or software.

It's an empire of **systems that think and leaders who care.**

9. The Lesson of DGIX

The greatest lesson I've learned from building DGIX is this:
When vision drives technology, the result is evolution.When technology drives vision, the result is entropy.

Every system must serve something higher than itself — a philosophy, a mission, a purpose.

DGIX exists to remind the world that the intersection of law, leadership, and intelligence is not a boundary — it is a blueprint.

And if we follow that blueprint, we will build not just smarter companies, but a smarter civilization.

In the next chapter, we will confront the question that every builder of the AI Empire must face:

How do we govern this power?

Because an empire without law becomes tyranny —
and intelligence without ethics becomes chaos.

Chapter 6
Law, Ethics, and the AI Governance Frontier

Empires rise through innovation.

They endure through **law.**

No matter how advanced our technology becomes, civilization collapses the moment power operates without accountability.
Artificial intelligence is now the most powerful force humanity has ever created — capable of shaping economies, elections, and even emotion.
The question is not whether we can govern AI.
It is whether we are wise enough to **govern ourselves through it.**

1. The Law Lags Behind the Code

Technology moves at the speed of innovation.
Law moves at the speed of consensus.

Between those two velocities lies the **governance gap** — a dangerous vacuum where disruption outpaces regulation.

In that vacuum, decisions once made by humans are now delegated to algorithms:

Who gets a loan.
Who gets hired.
Who gets watched.
Who gets believed.

Our traditional legal frameworks were never designed for self-learning systems.
Tort law can't assign blame to an algorithm.
Contract law can't negotiate with a neural network.
And constitutional law has no precedent for machines that make moral judgments.

The frontier of AI governance isn't about controlling machines — it's about **redefining accountability in the age of autonomy.**

2. The Code of Law vs. The Law of Code

As an attorney and technologist, I see a growing convergence:
Code is becoming law, and law must become code.

Every algorithm we design carries norms, rules, and power dynamics. When an AI moderates speech, it is exercising constitutional authority.

When it allocates healthcare resources, it's practicing policy.
When it sets bail or predicts recidivism, it's interpreting justice.

The architecture of algorithms has become the new **constitutional order** — invisible, unvoted, but profoundly influential.

That's why governance must evolve from written statutes to **encoded ethics.**

Future systems of law will not just be written in statutes, but programmed into systems. And the next generation of legal minds must be bilingual — fluent in both **legal language** and **machine language.**

3. Digital Constitutionalism

Every civilization needs a constitution — a foundational agreement on what is just, what is protected, and what is possible.

The AI Civilization needs the same — a **Digital Constitution** grounded in three principles:

1. **Transparency:** Every algorithm that affects a human life must be explainable.
2. Black-box systems have no place in a society that values accountability.
3. **Consent:** Individuals must own their data as they own their property.
4. Consent must be explicit, continuous, and revocable.

5. **Equity:** The benefits of intelligence must be distributed fairly — not concentrated among a technological elite.
6. Civilization cannot survive algorithmic aristocracy.

These principles form what I call **Algorithmic Rule of Law** — the belief that systems should be governed by predictable, auditable frameworks that uphold human dignity.

4. The Role of the General Counsel in the Age of AI

In the 20th century, General Counsel was a corporate guardian.
In the 21st, he is a **strategic architect of ethics.**

At **Disney Group International Exchange (DGIX)**, I redefined the role to extend beyond compliance.

Our legal team collaborates with engineers to encode ethical parameters directly into system design.

Before deployment, every AI model is reviewed not just for accuracy — but for integrity.

We ask three questions:

- Does it align with the client's mission?
- Does it respect the user's autonomy?
- Does it preserve transparency under pressure?

This integrated approach — what I call **Legal by Design** — transforms governance from oversight to architecture.
It's proactive, not punitive.
It's design thinking applied to justice.

5. The Ethics of Automation

Automation, without ethics, is exploitation at scale.
When AI replaces decision-making, it doesn't erase bias — it amplifies it.
When AI optimizes for efficiency, it often erases empathy.

We must remind ourselves that intelligence without empathy is **calculation**, not **consciousness.**

The ethical architect asks: What is this system optimizing for — profit, precision, or people? The answer determines whether AI becomes a servant or a sovereign.

At DGIX, we operate under the principle that **ethics must precede efficiency.** That's not moral sentiment — it's strategic necessity.
Empires built on exploitation always fall.
Empires built on ethics always evolve.

6. The Global Governance Imperative

AI governance cannot exist in silos.
Data flows across borders; so must accountability.

Just as the Geneva Convention established human rights for wartime, the coming decades will require an

Algorithmic Convention — a global framework for data rights, transparency, and digital sovereignty.

Imagine a world where every AI system, from Beijing to Berlin, must adhere to a shared set of moral standards — where machine learning models are audited like financial systems and certified like pharmaceuticals.

That's not utopian — it's **structural stability for civilization.**

7. Law as the Nervous System of Civilization

Law is civilization's nervous system.
It translates pain into protection, intent into action, justice into structure.

AI challenges that equilibrium — not by breaking laws, but by bending the boundaries of who or what is subject to them.
Do algorithms have liability?
Do data sets have rights?
Do digital agents have duty?

These questions may sound abstract — but they will soon determine the future of freedom.

Because in a world run by intelligent systems, **governance is no longer optional — it's existential.**

8. The Moral Algorithm

In the end, every empire must decide what it stands for.

The Roman Empire stood for order.
The British Empire for expansion.
The American Empire for liberty.
The AI Empire must stand for **wisdom.**

Our moral algorithm must optimize not for dominance, but for **dignity.** It must learn not just from data, but from conscience.

It must balance efficiency with empathy, speed with sovereignty, logic with law.

That is the challenge of our time — and the opportunity of our generation.

We are the first civilization in history capable of encoding ethics directly into the architecture of power.
If we succeed, the AI Empire will become the most enlightened system humanity has ever built.
If we fail, it will become the most efficient tyranny ever conceived.

The choice is ours.

And the code is being written now.

Chapter 7
The Human Element — Emotional Intelligence in a Machine World

Every empire has its engineers.
But the ones that endure have their **empaths.**

We have built systems capable of recognizing patterns, predicting outcomes, and imitating human reasoning — but we have yet to build one that can feel. Artificial intelligence can learn, but it cannot **care.** It can respond, but it cannot **relate.**

As our machines grow more intelligent, our greatest challenge will not be teaching them to think — it will be remembering how to be human.

1. The Rise of the Rational Machine

We are living in an age obsessed with optimization.
Efficiency has become the modern deity.
Every process, every platform, every person is measured by performance.

Yet, what makes us human has never been efficiency — it's empathy. The ability to sense beyond data, to feel beyond logic, and to act beyond self-interest.

The danger of the AI era is not that machines will become too smart, but that humans will become too mechanical.

A civilization built solely on logic will eventually collapse under its own precision.
Because systems without soul become **soulless systems** — perfectly efficient, but morally empty.

2. Emotional Intelligence as Strategic Capital

In the AI Empire, **emotional intelligence** is not a soft skill — it is a strategic asset. The leaders of tomorrow will be measured not only by what they know, but by how they connect — to people, to culture, and to conscience.

Empathy builds trust.
Trust builds cooperation.
And cooperation builds systems that scale sustainably.

At **Disney Group International Exchange (DGIX)**, we treat emotional intelligence as a pillar of enterprise design.

Every leadership meeting begins with three questions:

- What do our people need to feel empowered?
- What do our clients need to feel understood?
- What does society need to feel safe?

Because before you can engineer efficiency, you must engineer trust.

3. The Machine Learns the Mind, but Not the Heart

AI can analyze sentiment, but it cannot experience emotion.
It can simulate compassion, but not sincerity.

The subtlety of human emotion — love, guilt, pride, fear — cannot be replicated through code because it originates in consciousness, not computation.

Machines can understand "what" we feel, but not why.

That distinction is critical.
Because in the coming decades, as AI becomes capable of managing every measurable process, the only irreplaceable value left will be the **immeasurable human spirit.**

Leaders who cultivate emotional intelligence will not compete with machines — they will complement them.

4. Leading Humans in the Age of Algorithms

The AI era demands a new leadership model — one that balances analytics with empathy.
We call it **Symphonic Leadership** at DGIX: the art of orchestrating human and machine intelligence into harmony.

In a symphony, every instrument matters.

The violins cannot replace the percussion; the conductor cannot outplay the orchestra.

Likewise, in a hybrid organization, leaders must know when to trust data and when to trust intuition.

True leadership is not about being infallible — it's about being in touch.

The best leaders of the next century will be those who can interpret signals — both digital and emotional — and align them toward a shared vision.
That requires **emotional literacy**, **contextual awareness**, and **moral imagination.**

5. The Stoic Leader in the Digital Age

Stoicism, one of my guiding philosophies, teaches that power is not control — it is composure. In chaos, the Stoic does not react; he reflects.
In uncertainty, he does not panic; he prepares.
In success, he does not boast; he builds.

In a world of automation and volatility, this ancient discipline becomes modern leadership technology.

Emotional intelligence is not just about feeling deeply — it's about feeling deliberately. Stoicism transforms emotion into energy, and empathy into endurance.

Marcus Aurelius would have understood today's AI challenge perfectly:

"You have power over your mind — not outside events. Realize this, and you will find strength."

That is the mindset the AI age demands — leaders who

can regulate their own emotions before regulating machines that mirror them.

6. The New Hierarchy of Intelligence

In the past, IQ was the measure of intellect.
Today, AI owns IQ.

What remains uniquely human is **EQ — emotional intelligence**, and **CQ — consciousness quotient.**

- **IQ** measures what you know.
- **EQ** measures how you connect.
- **CQ** measures how you understand meaning.

AI can replicate IQ endlessly.

It will never replicate CQ — the uniquely human ability to create context,purpose, and values. That is why the next empire won't belong to the smartest systems, but to the wisest leaders.

7. Humanity as the Ultimate Interface

As interfaces evolve — from screens to voice, from voice to thought — the most powerful interface will always be the **human presence.** A handshake still builds more trust than a login.

A conversation still reveals more truth than a dashboard.

And a leader's integrity still carries more weight than any algorithm's output.

In every age of empire, what sustained civilization wasn't technology — it was connection. The AI Empire will be no different.

We must build systems that don't just replicate humanity, but reinforce it. Every algorithm should enhance human dignity, not diminish it.
Every advancement should bring us closer, not colder.

8. The Emotional Architecture of the AI Empire

At DGIX, we believe the next frontier of design is **emotional architecture** — the embedding of empathy, ethics, and intuition into the systems that shape our lives. It's not enough for AI to be accurate; it must also be appropriate. It must reflect not just intelligence, but intention.

The future will not be about "human vs. machine," but "human through machine."

AI will become the mirror through which we study ourselves — our patterns, our priorities, our prejudices.

If we are wise, we will use that reflection not to replicate humanity's flaws, but to refine humanity's form.

9. Legacy Through Humanity

The ultimate measure of an empire is not how high its towers rise, but how human its systems remain.

Technology is temporary; humanity is timeless.

We can replace every function of labor, but never the function of love.

We can automate decisions, but not decency.

As we march deeper into the machine age, our greatest duty is not to outthink the machines — but to outlove them.

That is the heart of The AI Empire:
to build a smarter world without losing our soul.
Because intelligence may build the empire,
but only **humanity** will make it worth inheriting.

Chapter 8
From Data to Destiny — The Economics of AI Empires

Every empire in history has been built on a scarce resource.
For Egypt, it was the Nile.
For Rome, it was trade.
For Britain, it was coal and colonies.
For the United States, it was oil and innovation.

But the resource that defines the twenty-first century is neither land nor labor.

It is **intelligence.**

Not just human intelligence — but organized, amplified, and artificial intelligence.

Data is the raw material. Algorithms are the machinery. Cloud systems are the factories.

And together, they form the economic engine of a new civilization — the **AI Empire.**

1. The Currency of Cognition

In traditional economics, wealth flowed from materials and markets.
In modern economics, wealth flows from **models and minds.**

The more intelligence your system can generate, the more value it creates.

That intelligence can take many forms — predictive analytics, autonomous logistics, generative creativity — but it all converts into one universal currency: **decision power.**

Every organization today competes on one question:

How fast and how accurately can you decide?

The AI-driven enterprise doesn't simply analyze the past; it **anticipates** the future.

Its balance sheet is no longer defined by assets, but by algorithms.

Its most valuable product is insight.

At **Disney Group International Exchange (DGIX)**, we call this model **Cognitive Capitalism** — an economy where information becomes infrastructure, and understanding becomes income.

2. The Shift from Labor to Learning

For centuries, economic value was measured by the productivity of hands.

Now it is measured by the productivity of systems.

AI has transformed labor economics into **learning economics**.

Machines can replicate tasks infinitely, but only systems that learn faster create compounding advantage.

The more your AI learns, the more efficient your entire organization becomes — not through scale alone, but through self-improvement.

That is why the most valuable companies in the world today — from NVIDIA to OpenAI to Oracle — don't sell products; they sell learning loops. They are not just factories of output, but factories of understanding.

3. Data as the New Land

In the feudal era, power was tied to land ownership.
In the AI era, it's tied to **data ownership.**

Every dataset is a digital territory — an ecosystem of patterns that can be mined, monetized, and weaponized.

The nations and corporations that control the largest and cleanest datasets control the modern equivalent of real estate.

But unlike land, data is infinite — and exploitable.

Its value depends on **governance.** Who owns it? Who can use it? Who protects it?

This is why DGIX treats data not as property, but as sover-

eignty. Just as constitutions protect citizens, data governance frameworks protect civilization.
Freedom in the AI era means controlling not just your labor, but your information identity.

4. Algorithms as Assets

In accounting terms, algorithms are intangible.
In reality, they are institutional.

An algorithm that predicts demand better than competitors can shift entire markets.

An algorithm that optimizes logistics can collapse costs and dominate industries.

The new monopolies will not own pipelines or ports — they will own patterns.

As a legal technologist, I believe we must recognize algorithms as **economic actors** — subject to accountability, taxation, and ethical audit.
Because in the global AI economy, algorithms aren't just tools; they are traders.

4. Algorithms as Assets

In accounting terms, algorithms are intangible.
In reality, they are institutional.

An algorithm that predicts demand better than competitors can shift entire markets.

An algorithm that optimizes logistics can collapse costs and dominate industries.

The new monopolies will not own pipelines or ports — they will own patterns.

As a legal technologist, I believe we must recognize algorithms as **economic actors** — subject to accountability, taxation, and ethical audit.
Because in the global AI economy, algorithms aren't just tools; they are traders.

6. The Intelligence Supply Chain

Just as industrial empires required supply chains of steel and oil, AI empires require supply chains of data and compute.

At DGIX, we map intelligence supply chains in four layers:

1. **Data Extraction** - Collecting and classifying meaningful signals.
2. **Data Refinement** - Cleaning, labeling, and securing information.
3. **Model Construction** - Training and deploying algorithms that interpret the refined data.
4. **Insight Distribution** - Delivering intelligence to the right human decision-makers.

Each layer generates its own economy — analysts, architects, and auditors.

Together, they form the **engine of exponential value.**

7. Economics of Trust

No empire can function without trust — not political, and not digital.
In the AI economy, trust is both currency and collateral.

Companies that use data unethically may gain temporary advantage, but they lose credibility capital — and in the long term, that is fatal.

Sustainable profitability in the AI era depends on transparent governance, responsible AI, and auditable systems.

At DGIX, we tell our clients:

"Data gives you power. Governance gives you permanence."

Without ethical architecture, efficiency becomes exploitation; intelligence becomes intrusion.
Trust transforms both into progress.

8. The Destiny Equation

Economic destiny in the AI age can be summarized by one formula:

Vision × Systems × Integrity = Longevity.

- Vision sets the direction.
- Systems deliver the scale.
- Integrity ensures survival.

This is the equation that built DGIX, and it's the same equation that will define the future of every enterprise, nation, and civilization.

Because data without vision is noise.
Systems without integrity are fragile.
And intelligence without ethics is chaos disguised as progress.

9. Building a Smarter World Economy

The true promise of AI is not to create trillion-dollar valuations — it's to create **intelligent value chains** that reduce waste, empower creativity, and extend opportunity.

An AI-driven economy can heal supply chains, democratize knowledge, and reimagine labor itself.

It can make work more meaningful and societies more adaptive — if guided by design and discipline.

That is the purpose of The AI Empire: to show that intelligence is not only power, but potential. And the destiny of civilization is not determined by who builds the biggest models, but by who builds the **wisest systems.**

The next chapter will explore exactly that — how leaders ensure their systems endure.

Empires rise on intelligence; they last on **legacy.**

Chapter 9
The Legacy Equation — Vision × Systems = Immortality

Every empire begins with ambition.
But only the wise ones are built for eternity.

History remembers Rome not for its conquests, but for its code of law.
It remembers Greece not for its wars, but for its wisdom.
It remembers the Renaissance not for its wealth, but for its wonder.

Empires rise by innovation — but they endure by institution.
And in the age of artificial intelligence, the new measure of greatness will not be how much data you own, but how deeply your systems endure.

1. The Mathematics of Legacy

Legacy is not memory — it is a mechanism. It is what continues to operate long after its originator is gone.

Every lasting institution — from the Vatican to NASA to Oracle — has survived because it transformed **vision into system.**

That's the legacy equation:

Vision × Systems = Immortality.

Vision gives meaning.
Systems give continuity.
Together, they produce permanence.

Without vision, systems decay.
Without systems, vision dies.
But when they align — when structure serves purpose — an empire achieves what every human soul quietly desires: **to outlast itself.**

2. From Leadership to Architecture

Most leaders think in terms of goals.
Architects think in terms of generations.

The difference is profound.

A goal seeks achievement; architecture seeks alignment.

A leader manages; an architect designs.

Leadership wins quarters. Architecture wins centuries.

At Disney Group International Exchange (DGIX), we train leaders to think architecturally — to build frameworks, not just companies; to construct ecosystems that can self-sustain through clarity, accountability, and culture.

We teach them that true leadership is not control — it is continuity.

3. Systems as Memory

Every system is a form of memory.

A process remembers what people forget.

A protocol preserves what passion fades.

That is why system design is an act of **preservation.** When we design workflows, governance models, and knowledge networks, we are encoding culture — ensuring that purpose survives turnover, trend, and time.

The pyramids were physical systems of faith.
Constitutions are legal systems of belief.
AI architectures are digital systems of knowledge.

The medium evolves; the motive does not.
All systems exist to make meaning immortal.

4. DGIX: Building for Permanence

DGIX was never designed as a company — it was designed as a **continuum.** A living ecosystem of principles, technologies, and people aligned around one ideal: to build responsibly and intelligently.

Our methodology ensures that every engagement, every product, every training program contributes to a shared infrastructure of intelligence.

We don't just deliver projects; we **build permanence.**

Our Oracle ERP integrations teach scalability.
Our AI governance frameworks teach accountability.
Our thought leadership — in law, leadership, and systems — teaches philosophy.

DGIX isn't merely scaling business; it's **scaling legacy.**

5. The Philosophy of Immortality

Every great builder eventually confronts a question:

"How do I build something that lives beyond me?"

The answer lies not in replication, but in **design with purpose.** When you embed principle into process, your values become structural.

When you embed culture into code, your leadership becomes algorithmic.

When you embed wisdom into systems, your life becomes immortal.

That is how civilizations endure — by encoding their best selves into their infrastructures.

6. Institutional Integrity

Empires rarely fall from external attack; they decay from **internal entropy.** Corruption, confusion, and complacency are the three viruses of legacy.

The only antidote is integrity.

Integrity is the connective tissue of all enduring systems.

It ensures that vision remains uncorrupted, even when leadership changes hands.
That's why DGIX prioritizes governance design: every system we build has embedded transparency, accountability, and adaptability.

An empire protected by armies will survive decades.

An empire protected by integrity will survive centuries.

7. The Legacy Leader

The modern empire builder must think like a philosopher and act like an engineer.

He must balance ambition with accountability, creativity with control, innovation with introspection.

Legacy leaders understand that the systems they build today are the culture their successors will inherit tomorrow.

They design not just for performance, but for permanence.

The goal is no longer to be indispensable — it is to design systems so complete that your absence becomes proof of your leadership.

8. The Immortality of Ideas

Technology changes; ideas persist.
Systems evolve; principles endure.

Marcus Aurelius' thoughts outlived Rome.
Tesla's visions outlived his patents.

Steve Jobs' design language outlived his era.

The AI Empire will be defined not by the size of its models, but by the meaning of its mission. Empires built on innovation are impressive.

Empires built on **ideals** are immortal.

9. Legacy as a Living System

Legacy is not a monument — it's a **mechanism that moves forward.** It adapts, learns, and grows — like an organism designed for eternity.

That is how we must think of every organization we build in the AI age.
Your company is not a corporation; it's a living system.
Your technology is not just infrastructure; it's inheritance.

And your legacy is not what you leave behind — it's what you build into everything that lasts.

We build systems so our wisdom does not die with us.
We design structures so our intentions survive scrutiny.
We codify ideas so our visions remain executable.

Because in the end, immortality is not about living forever
—
It's about building something that does.

That is the destiny of The AI Empire:
To design systems that think, serve, and **endure.**

Chapter 10
The Future We Are Building

Every age believes it stands at the summit of progress.

But only a few ages have truly built the future.

The industrial age built with steel.
The digital age built with code.
Our age — the age of intelligence — builds with consciousness.

We are constructing not just smarter machines, but a smarter world — one where intelligence itself becomes the raw material of civilization.

But the question remains: what kind of civilization will it be?

1. The Moment Before the Next Epoch

History always pivots quietly.

No trumpet announces the turning of an age.

Right now, as billions of algorithms hum invisibly beneath our lives, humanity stands at one such pivot.

This is not the era of gadgets — it is the **genesis of governance by intelligence.** Every click, contract, and computation contributes to the architecture of tomorrow.

The future is not something that happens to us; it is something we **design into existence.**

And that design depends not on code alone, but on **character.**

2. Technology as Reflection

Artificial intelligence is not our replacement — it is our reflection. It reveals who we are, what we value, and what we fear.

It mirrors our curiosity, but also our contradictions.

And through that reflection, we have the rare opportunity to edit humanity itself — not biologically, but **behaviorally.**

Technology doesn't define morality.
It exposes it.
It forces us to confront what kind of beings we are when power becomes programmable.

That is why The AI Empire is not a book about machines — it is a book about **meaning.**

3. The Architects of Tomorrow

The next generation of leaders will not rule with armies or ideologies.

They will lead with **architecture.** They will design systems that harmonize intelligence and intention.

They will be philosophers with dashboards, engineers with ethics, and entrepreneurs with empathy.

They will understand that an empire's power lies not in its algorithms, but in its alignment — the moral coherence between what it builds and what it believes.

At **Disney Group International Exchange (DGIX)**, this is our creed:
We don't build to impress. We build to improve.

Every model, every framework, every strategy must move civilization forward — even by one principle, one innovation, one life at a time.

4. Governance as the New Greatness

In ancient times, greatness was measured by conquest.
In modern times, it was measured by consumption.
In the age of AI, greatness will be measured by **governance.**

The leaders who will define this century are those who can channel intelligence into justice — who can use AI not merely to accelerate business, but to elevate being.

They will rewrite the contract between humanity and technology — ensuring that power serves purpose, and not the reverse.

The AI Empire must not be an empire over people.

It must be an empire for people.

5. From Systems to Souls

The more intelligent our systems become, the more intentional our souls must be.
Because automation without awareness leads to emptiness.
We can automate every decision — but not discernment.
We can replicate every thought — but not wisdom.

The future belongs to those who build with empathy as rigorously as they code with precision.

That is the great paradox of progress:

The smarter our machines become, the more **moral** our leadership must be.

6. The Symbiosis of Human and Machine

The ultimate goal of AI is not to surpass humanity, but to complete it. To extend our reach, deepen our insight, and amplify our imagination.

When guided by integrity, AI becomes the greatest mirror of human potential.

When corrupted by ego, it becomes the greatest multiplier of human error.

The path forward is not domination or dependence, but **symbiosis.** A world where human wisdom and machine intelligence operate as partners — one providing conscience, the other capacity.

That is the civilization worth building — one where intelligence, in all its forms, exists in harmony.

7. The Responsibility of the Builders

The architects of this age — engineers, policymakers, educators, entrepreneurs — carry the same weight once borne by lawgivers and philosophers.

We are designing the invisible infrastructure of human destiny.

To build systems without ethics is to build bridges without foundations.

To program intelligence without humility is to summon power without purpose.

Every line of code, every corporate charter, every policy decision now contributes to the same architecture: **the operating system of civilization.**

This moment demands courage — not the courage to dominate, but the courage to design rightly.

8. The DGIX Principle: From Local Systems to Global Consciousness

From our headquarters in Columbus, Ohio, **DGIX** began as a vision — a belief that local intelligence can shape global consciousness.

That a small team with a big philosophy can influence how the world builds.

Today, as we collaborate across industries and borders, I'm reminded that empires are not defined by scale — they're defined by **significance.** A single idea, when encoded into the right system, can alter the trajectory of history.

That's the essence of what we call the DGIX Principle:

"When you design intelligently, you design indefinitely."

Every decision — from architecture to ethics — ripples into eternity.

9. The Age of Intelligent Civilization

We are not merely entering the Fourth Industrial Revolution.
We are entering the **First Intelligent Civilization.**

A civilization where consciousness becomes capital.
Where empathy becomes efficiency.
Where law becomes logic, and logic becomes law.

And in this new order, the empire that leads will not be the one that conquers the most markets — but the one that cultivates the most mindfulness.

The great race of our time is not between nations or corporations — it's between our **capacity to create** and our **courage to care.**

10. The Final Vision

The story of The AI Empire is not about me, or DGIX, or technology.

It is about all of us — the architects, the dreamers, the disciplined minds shaping destiny through design.

We have been given a rare gift: the ability to build intelligence itself.

The question now is whether we will build it wisely.

If we lead with systems alone, we will build machines that outlast us.

If we lead with vision and virtue, we will build a **civilization that uplifts us.**

That is the future we are building.

A future defined not by domination, but by design.

Not by control, but by conscience.

Not by how intelligent our systems become — but by how human we remain.

Epilogue

The empire rises quietly — not through conquest, but through connection.
Through vision turned into structure.
Through structure turned into legacy.
Through legacy turned into **light.**

And one day, when history looks back at this moment — it will not remember the names of the companies or the machines.

It will remember that humanity, at the height of its power, chose to build **wisely.**

Because the true empire — the only one that endures — is the one built with systems and **vision.**

Appendix A: Influences and Recommended Reading

This book draws on philosophy, systems theory, economics, law, and emerging technology. The following works are offered as references for readers who wish to explore the ideas behind each chapter more deeply. Each section aligns directly with the themes discussed in the corresponding chapters.

Chapters 2 and 4: Systems Thinking and Architecture

These chapters explore the idea that "Architecture is destiny" and why leaders must think like "Systems Architects".

- ***Thinking in Systems: A Primer* by Donella Meadows.**
  - This is the foundational introduction to the "System Mind". It explains how feedback loops and leverage points work, which supports the

author's view that leaders must manage "patterns, not people".
- *Clean Architecture* **by Robert C. Martin.**
    - Especially valuable for technical readers, this book reinforces the author's point that "structure determines behavior" and that software must be built to be independent of frameworks.
- ***"Conway's Law"*** **(Melvin Conway).**
    - Conway's insight supports the argument in Chapter 4 regarding "The Law of Interconnectedness". It posits that systems inevitably reflect the communication structures of the organizations that build them.

Chapters 3 and 6: Law, Ethics, and Governance

These chapters call for a *Digital Constitution* and an *Algorithmic Rule of Law*, arguing that power at digital scale requires enforceable rights, clear limits, and accountable oversight.

- ***The Age of AI: And Our Human Future*** **by Henry Kissinger, Eric Schmidt, and Daniel Huttenlocher.**
    - This book strengthens the "Empire" theme. It discusses AI as a geopolitical force and the need for new philosophical frameworks, mirroring the author's call for "Civilization by Design".
- ***Weapons of Math Destruction*** **by Cathy O'Neil.**
    - A necessary counterweight that grounds the ethical warning in concrete evidence. It

provides the evidence for the author's warning that "Automation, without ethics, is exploitation at scale" , specifically regarding loans and recidivism models.

- ***The EU Artificial Intelligence Act (EU AI Act).***
  - A major governance reference point and a concrete example of regulation catching up to reality. The world's first comprehensive AI law that claims the "Governance Gap" is being closed and aligns with the author's "Global Governance Imperative".

Chapters 7 and 10: Stoicism and The Human Element

These chapters argue that emotional intelligence is a strategic asset and draw on Stoic philosophy to explore the role of restraint, self-mastery, and human judgment in leadership.

- ***Meditations* by Marcus Aurelius.**
  - The primary source for the concept of the "Stoic Leader". This work reinforces the concept that power requires restraint and responsibility for one's inner state.
- ***Human Compatible* by Stuart Russell.**
  - Russell proposes designing AI systems that remain uncertain of their objectives so it defers to humans. This provides a technical foundation for the author's "Symphonic Leadership" and human-machine "Symbiosis".

- ***"How to tame the tech titans"* by Shoshana Zuboff (TED Talk)**
    - Zuboff's discussion of "Surveillance Capitalism" serves as a cautionary contrast to the author's notion of Cognitive Capitalism". It serves as a warning of what happens when "Data as the New Land" is not governed by ethics.

Chapter 8: The Economics of AI

The following works deepen the conceptual framework introduced in this chapter, particularly the ideas of the **Intelligence Supply Chain** and **Data as the New Land**. Together, they situate artificial intelligence within a broader economic transition - from industrial power to cognitive and perceptual power.

- ***The Second Machine Age* by Erik Brynjolfsson and Andrew McAfee.**
    - A foundational text on the transitionfrom industrial power (steam/steel) to cognitive power, directly supporting the "From Power to Perception" shift.
- ***"Data is the New Oil"* (Clive Humby) vs. *"Data is Labor"* (Glen Weyl).**
    - These perspectives illuminate the economic logic behind "Data Refinement". Weyl's arguments specifically support the author's view on "Sovereignty" and individuals owning their data.

www.ingramcontent.com/pod-product-compliance
Lightning Source LLC
LaVergne TN
LVHW022016200926
848481LV00007B/2428
* 9 7 9 8 9 0 2 2 2 8 5 9 2 *